Hi
Hi
Hi
Hi

MINNESOTA
CHILDREN'S PRESS
WWW.MINNCHILDPRESS.ORG

The mission of **Minnesota Children's Press** is to mentor rural students' writing and illustration skills, ages 5-15, for the purpose of revealing and amplifying voices of those seldom heard in society—yet who bear the full, long burden of the future that adults make today: young children. Through children's inborn creative caring, exuberant problem solving, and heroic imaginings, Minnesota Children's Press co-creates joyful, positive stories of community to unite and guide us to a generative future.

Designing, editing, and preparing this manuscript for publication was made possible by the generosity of many donors to Minnesota Children's Press, 501(c)(3) non-profit charity, including major support from memorials to the late Joyce and Gordon Lindquist of Grand Marais. We thank you all!

All Minnesota Children's Press Books are written, illustrated, published, and printed in the United States.

Second edition, revised. First published as a free, classroom teaching edition
by Minnesota Children's Press, 2024

10 9 8 7 6 5 4 3 2

This book was typeset in Fillip; titles in Chelsea Market Pro.
Illustrations were done in mixed media, with digital editing and enhancements.

ISBN: 978-1-7355753-7-7 • Library of Congress Control Number: 2024915575

GO AWAY, ROCK SNOT!

By STORY SCOUTS
of Grand Marais,
Minnesota, USA

A CHILDREN'S FIELD GUIDE

to the latest invasive species threatening North Shore streams: ROCK SNOT!

MINNESOTA
CHILDREN'S PRESS

Art by Ethan A.

ABOUT STORY SCOUTS

Story Scouts is an entrepreneurial publishing club for rural Minnesota kids ages 5-15 who want to write, illustrate, and publish their own books, newspapers, comics, graphic novels, websites, blogs, and more. Story Scouts create and publish limited editions, such as this book, of classroom demonstration projects that model collaborative group writing and illustration by young children. Classroom editions are free for the purpose of teaching applied literacy in creative ways. Story Scouts also sell their published products as fundraisers to support community projects they identify, while also building personal portfolios to be college- and career-ready.

Mentored by professional authors, designers, artists, editors, scientists, and other content experts, Story Scouts learn practical and artistic skills and earn achievement badges for demonstrating 21st century communication competencies.

CONTRIBUTORS

STORY SCOUTS

Adalyn	Jocelyn
Anson	Leah
Aren	Linnea
Ari	Madison
Brisa	Maite
Brynn	Mavis
Dax	Mikey
Elisa	Nathan
Ella P.	Nolan
Emily	Quinn
Emma	Rilynn
Ethan A.	Rosalyn
Ethan G.	Runa
Freddie	Sabrina
Freja	Selah
George	Sam
Grace	Sig
Iona	Tib
Jase	Tessa

EDITOR-IN-CHIEF, PUBLISHER, & WRITING MENTOR
Anne Brataas, Founder of Minnesota Children's Press

STICK FIGURE ARTIST & ILLUSTRATION MENTOR
Kip Hathaway

SENIOR EDITOR, BOOK DESIGNER, & DIGITAL ART ENHANCER
Brittany Lynk, BLynk Creations

ASSISTANT DESIGNERS AND DIGITAL ART ENHANCERS
Allison Healy, Rose Hathaway, and April Clark

SPECIAL THANKS FOR OUTSTANDING COMMUNITY HELP

★ Minnesota Department of Natural Resources
★ Science Museum of Minnesota
★ St. Croix Watershed Research Station
★ University of Minnesota-Duluth
★ Students attending Minnesota Children's Press traveling Comma Club Proofreading Workshop at Great Expectations Charter School, Grand Marais, MN
★ Support for our scientific mentors' field studies from the Legislative-Citizen Commission on Minnesota Resources (LCCMR), Minnesota Sea Grant, and Environment and Natural Resources Trust Fund (ENRTF).

TABLE OF CONTENTS

INTRODUCTION
WHY + HOW WE DID IT

Welcome to the worrisome world of rock snot!

Gross, right? Figure 1 shows a rock snot sample collected in the summer of 2024. Just looking at it you can see why no one wants shorelines of our pristine northern waters to be covered with this slimy algae.

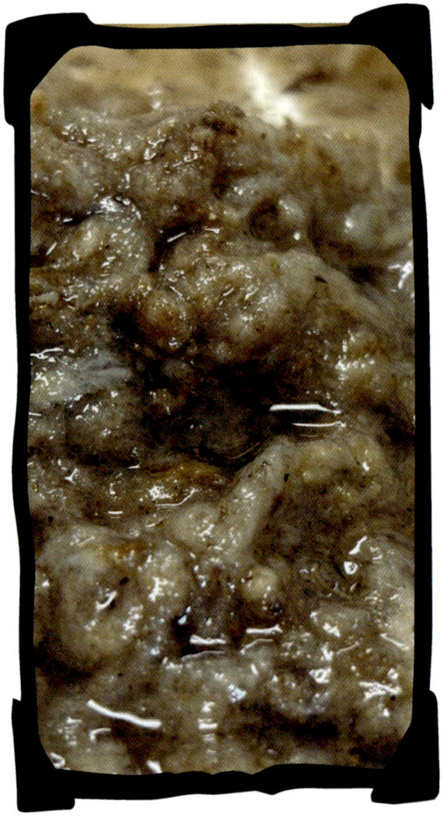

Figure 1: Eww! Rock snot!

Scientists are acting now, at the first signs of detecting it in our North Shore streams, rivers, and lakes, to teach the public about rock snot. That way we can all help prevent rock snot from spreading and making our beautiful freshwater too gross to wade, swim, boat, and fish in.

Our book is a true, evidence-based science story about the most recently documented invasive species threat to Minnesota's northern waters called *Didymosphenia geminata*—didymo, or rock snot, for short.

Photo credit: Science Museum of Minnesota

Why Research Rock Snot?

Scientists study rock snot to understand it so they can prevent rock snot from:

- lowering water quality
- causing trout to relocate due to lack of food
- destroying the natural beauty and bounty that humans rely on for work and play: tourism, fishing, swimming, boating, camping, nature retreats, artistic inspiration, spiritual renewal

Rock snot grows in colonies that cover up stream bottoms. Their stalks smother insects living there, which trout eat.

If trout can't find a good food source, they will swim to new waters to find more of their preferred trout food—stoneflies, mayflies, and caddisflies. Or, if they stay, fish will likely be smaller in size and fewer in number.

Mayfly

Stonefly

Caddisfly

Art by Kip Hathaway

True Science, With a Dash of Puppies

We wrapped this true science story in a made-up story about puppy scientists from Sunny Shores Puppy Preschool. We set the school and our story in Grand Marais, Minnesota.

In our book, puppies imagine a solution to controlling the threat that rock snot poses to northern waters. Yay for puppies!

About Us + Bottom-up Books

Story Scouts learning about rock snot from Dr. Mark Edlund of the Science Museum of Minnesota and Sarah Clauss of the University of Minnesota–Duluth.

We are 1st through 5th graders in schools of Cook County, Minnesota. For two summers we have collaborated on water sampling field work with scientists who came to Grand Marais from the St. Croix Watershed Research Station of the Science Museum of Minnesota. They are part of a team of researchers that also includes scientists from the Minnesota Department of Natural Resources and University of Minnesota-Duluth who have been studying rock snot for six years.

Dog art by Jocelyn, Brynn, and Mavis; Photo credit: Anne Brataas

During the school year, we worked after class in our Story Scouts Children's Publishing Club to write, draw, dictate, brainstorm, and act out rock snot science through skits.

Story Scouts acting out "Rock Snot"

Besides scientists, we were mentored by an early-career cartoonist and a few professional writers, copy editors, and designers to help combine words and images into this book. Story Scouts' Comma Club, whose members are dedicated to correct punctuation, helped proofread the book and performed the ritual last look for errors before it went to press.

Story Scouts Comma Club

This bottom-up, mentored editorial model of book production is the novel creation and hallmark of the Story Scouts Children's Publishing Club (www.storyscouts.org) of the 501(c)3 non-profit organization, Minnesota Children's Press (www.minnchildpress.org) based in Grand Marais.

MINNESOTA CHILDREN'S PRESS
WWW.MINNCHILDPRESS.ORG

Photo credits: Anne Brataas

Fixed It! Improving Reality with Rock Snotters

Our story improves reality. With help from the imaginations of our child-authors and -illustrators, the puppies discover predators we invented and named Rock Snotters. They eat rock snot to clean up lakes, rivers, and streams.

Rock Snotters do NOT exist. We wish they did.

In fact, no predators have yet been identified that could help control the spread of rock snot. Scientists are using their data, expert field research skills, and laboratory analysis to understand and control rock snot—and we are helping them*!*

Join us*!* Help teach everyone about rock snot to protect fish, insects, and our northern waters*! !*

Book cover designs

Art by Jase (Rock Snotters), Sig, Grace and Mikey, and Runa (bottom: L-R book cover designs)

Why We Protect Water

Our goal is to spark awareness about how to stop a rock snot invasion that could degrade water quality and crowd out the food sources in streams that fish need to thrive.

3 More Reasons to Protect Water from Rock Snot

It is wrong and wasteful to degrade water. Degraded water habitats are:

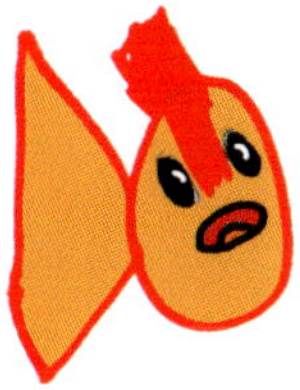

1. no longer healthy environments for the many creatures who live there or need its life-giving support.

2. not beautiful or restorative to the human spirit. The original Indigenous stewards of northern lands and waters have lived with Lake Superior and its tributaries for thousands of years, ever since the retreat of the last glaciers. We must continue to honor and protect these treasured natural resources.

3. economically damaging. Thousands of residents earn their living from healthy water. About 1.25 million visitors come to Cook County every year and most interact with our pristine waters in some way—generating $220 million annually in economic activity.

Art by Rilynn (all fish), Grace (rainbow), and Mikey (boat)

We Learned This

We all need to clean and dry our gear—boots, boats, fishing tackle—every time we leave one body of water, and before we enter another. If we don't, we risk tracking in bits of rock snot that can spread new growth. Right now, human behavior is the only control method for stopping rock snot. **Clean your gear by scrubbing it in hot soapy water after you leave a body of water and before you enter another.**

Scrubba-dub-dub your boots and gear.

ROCK SNOT PRO TIP: Pack a jug of hot water with dish soap in it, a scrubber, and a bin to catch the gear bath water. When finished, pour the gear bath water back in the jug and take it home to water the garden!

Art by Madison

Gratitude

This project was made possible through major funding from memorials of two Grand Marais residents, the late Joyce Russell Lindquist, who died in 2023, and her husband, Gordon Lindquist, who died in 2022.

Joyce and Gordon Lindquist

Joyce was Cook County's first kindergarten teacher, starting in 1953. She and Gordon were lifelong champions of children, families, reading, Cook County, fishing, and protecting Lake Superior and its watershed.

We also thank Library Friends of Cook County, Minnesota, and the many private donations who helped fund teaching space and supplies for working with scientists. We are deeply grateful for your support!

With appreciation for your reading, thinking, and caring,

Anne

Anne Brataas, Editor-in-Chief & Publisher
Story Scouts of Cook County and
Founder of Minnesota Children's Press

Gordon Lindquist with a 37-pound lake trout he caught in 1948 off Grand Marais harbor

Photo credits: (top) Julie Lindquist Lehmann; (bottom) Photo courtesy of Cook County History Museum

PART 1:

OUR SCHOOL

One day, we puppies of Sunny Shores Puppy Preschool were happily playing "hide-the-bone-bandana tag" on the playground, racing around hiding the bone, finding the bone, and then tearing off the bandana of the "it" puppy to tag them out.

Then we heard an announcement:

Your new substitute teacher has arrived!

We ran inside!

Art by Leah (dog)

The substitute dressed like a human. But something about her told us she might be a dog wearing a human costume. The costume even covered her paw with a rubbery hand that had a tattoo of a giant snake, with its tongue out, tasting her finger. Its scaly body wrapped around her arm. On her thumb, there was a tattoo of a cube with dots on it.

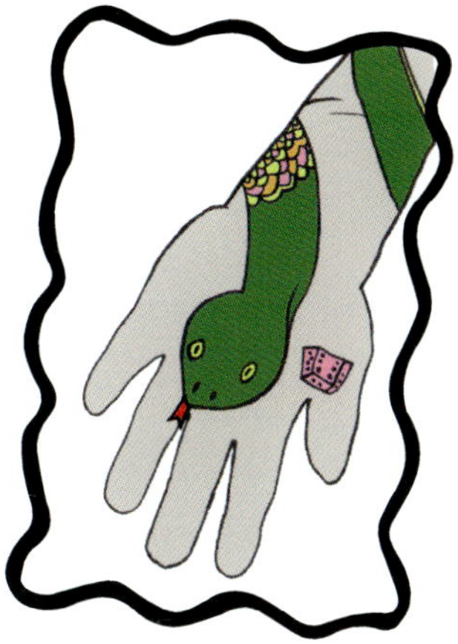

Very interesting.

Her name was interesting, too. **Dr. Definitely, Totally Not Evil**—or Dr. Definitely, for short.

Compared to her, our names were easy:
Sir Oshibon Bingsnucker, Bob, Bob-O, Bob-E, Bobo, Mia, Lilac the dalmatian, and Puppy Prof, the genius puppy who graduated from Puppy University before she lost all her puppy teeth.

Art by Tessa (Dr. Definitely), Brynn (hand), and Anson (Puppy Prof)

We told Dr. Definitely that our regular teacher always takes us on walks to learn about the world because this is what you do in Puppy Preschool.

But on this one day, because the world is so very interesting, we forgot our Puppies of Sunny Shores Pledge:

SUNNY SHORES PUPPY PRESCHOOL PLEDGE

I will listen to my teachers, substitute teachers, and grown-ups who help me learn.

Art by Mavis (dogs)

It's possible we didn't hear her because we were so happy, we were running around in circles.

Circles are the best!

Or, we thought she was talking to other puppies.

Dog art by Jocelyn, Runa, Sabrina, and Tib

We also thought that since she was so unusual it would be OK to tell her something unusual, and not exactly correct: *that our regular teacher always let us swim in the streams near school!*

Actually, our regular teacher did this once—we got to swim in Triplet Streams.

We puppies were so busy running circles of happiness we didn't hear Dr. Definitely's warning:

Art by Brynn (Triplet Streams), Jocelyn (dogs), and Freja (golden unicorn)

Art by Tessa (Dr. Definitely) and Emma (puppies)

PART 2:

OUR PROBLEM: Rock Snot

We learned why Dr. Definitely warned us when we took a quick detour on the way to the Triplet Streams. We stopped into one of our favorite stores, **Hackers & Hoodies,** to see what was on the computer screens in the window.

On the main computer screen, we watched a video explaining how rocks can have gloppy, gross things stuck to them called didymo, or **ROCK SNOT** for short! For long, you can call it by its scientific name: **DIDYMOSPHENIA GEMINATA.**

Art by Anson (Hackers & Hoodies store logo and decor)

Rock snot is an invasive species, a kind of algae called a **DIATOM**. It pollutes streams by making thick gloppy mats that smother things like **INSECT LARVAE** that fish like trout love to eat and need to survive.

Photo credits: Science Museum of Minnesota; Art by Kip Hathaway

Gloppy mats of rock snot are easy to see. But the tiny parts that spread rock snot are invisible unless you have a **FANCY MICROSCOPE.**

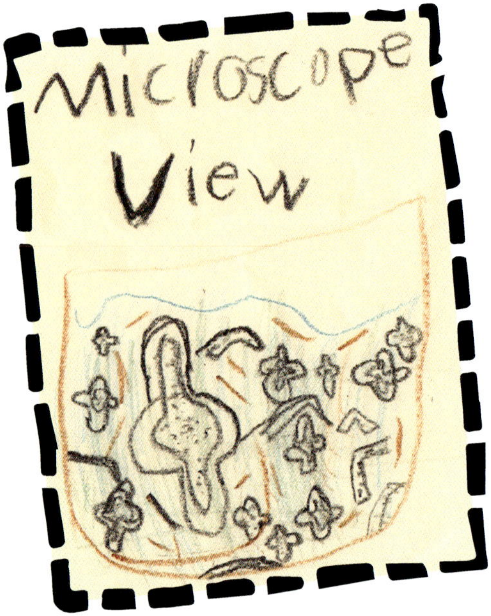

FANCY MICROSCOPES

Fancy microscopes magnify a single cell of rock snot so it can be seen 1000 times bigger on a computer screen!

Art by Jocelyn (microscope); Photo credits: Anne Brataas

PETRI DISH DISCOVERIES

A single rock snot cell is 0.1 mm long.

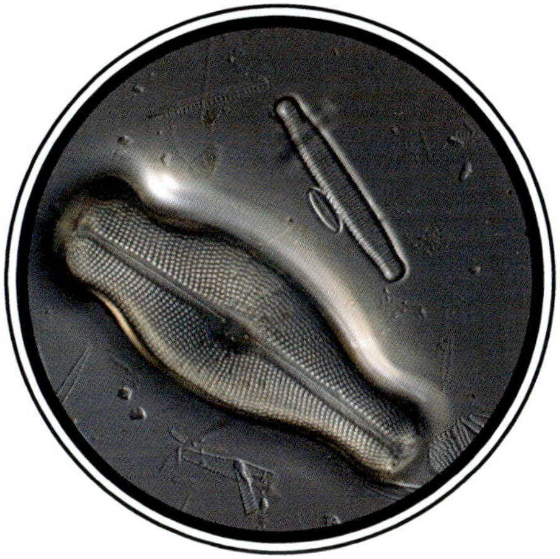

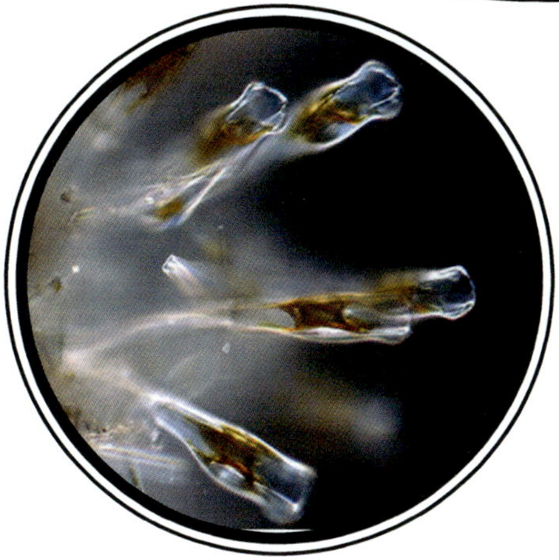

Rock snot cells and stalks magnified 2,000x

Art by Nolan (dog); Photo credits: Science Museum of Minnesota

The Hackers & Hoodies store manager printed out a bunch of rock snot pictures from the video. We learned to be careful not to track tiny, hitchhiking pieces of rock snot clinging to fur, waders, boats, or fishing gear from a snotty stream to a clean stream.

Photo credits: Science Museum of Minnesota; Art by Linnea (dog)

He tucked the rock snot pictures into hoodie pouch pockets. Then he gave us each a free Hackers & Hoodies hoodie for helping him get the word out about how to stop the snot.

We headed straight for the Triplet Streams and splashed right in!

Somewhere in the distance it sounded like someone was saying,

**"Oh c-l-a-a-a-s-s-s-s!
Here Puppies!
Who's a good puppy?!!!
Here puppy, puppy, puppy!!!"**

It must be that video at Hackers & Hoodies.
Hope we get to watch it on the way home!

Art by Leah (dog)

PART 3:

OUR SCIENCE SOLUTION: Fish School

While we were swimming, playing underwater, and collecting agates, we met a giant mother fish. She was very friendly and welcoming.

We discovered we could speak to each other in Bubble language!

Art by Rilynn (school of fish), Linnea (dog and fish), and Emma (Mother Fish's world)

Speaking slowly in Bubble, Fish Mom told us she started a school because she was worried about rock snot covering rocks and smothering all the fish food hiding around the rocks. That could harm fish!

She said the rocky stream bottom is a special habitat for raising baby **INSECTS** called **LARVAE**. They are important because fish eat them to grow and thrive.

Art by Mikey (underwater world), Leah (talking pups), and Kip Hathaway (underwater creatures)

To protect the stream habitat, Fish Mom teaches all watery things about rock snot and how to stop the spread.

STOP THE SNOT PLEDGE

I will always clean my things after leaving one body of water and before entering another.

Keep your boots and fins sparkly clean.

Scrub your gear with hot, soapy water!

She asked us:

"Would you puppies like to go to a class about rock snot in Fish School?"

"YES YES YES!!!" we Bubbled together!

Art by Madison (boot), Mikey (boat), Leah (talking pups), and Linnea (Mother Fish)

"Then follow me," she said with a swish of her gleaming tail. We puppy-paddled behind her into a beautiful rainbow cave on the bank of the stream.

It was perfect*!* We puppies could easily come up for breaths of air in the upper cave, then go down underwater to join the fish students in the lower cave.

Art by Grace (rainbow cave), Emma (fish background), Linnea (red fish, brown dog), and Jocelyn (purple dog, green fish)

Class had just started. The lesson was showing a video Fish Mom found on a scientist's computer tablet that fell out of a boat while the scientist taught children about rock snot.

When the tablet fell, several fish managed to drag the rubber-framed glowing tablet into the cave to hide it. We knew how we could help: Puppy guard it! We surrounded it in a blur of beating tails as a scientist's hand plunged into the water to find the tablet.

Suddenly the water was calm. The hand stopped thrashing—and disappeared!

The scientists gave up, and the tablet was ours!

Art by Rilynn (underwater world), Brynn (reaching for tablet), and Grace (rainbow cave)

We were so happy because we got to watch the tablet's rock snot video! Even though we chatted in Bubble with our new fish friends all through the video—which was named "Q & A"—we still learned important science from it.

Here are our favorite parts of "Q & A":

Art by Leah (puppies); Photo credit: Miles Monson, Kui Hu, and Mark Edlund from Science Museum of Minnesota

Q: WHAT IS ROCK SNOT?

A: Rock snot is invasive cold-water algae that grow in big, thick mats or clumps called colonies. It is a specific kind of algae, a diatom, and it has a long name: *Didymosphenia geminata.*

Call it "didymo" for short. And rock snot for fun!

Rock snot is a nickname. It got the nickname because it looks gloppy, slimy, and gross—like it got sneezed onto rocks.

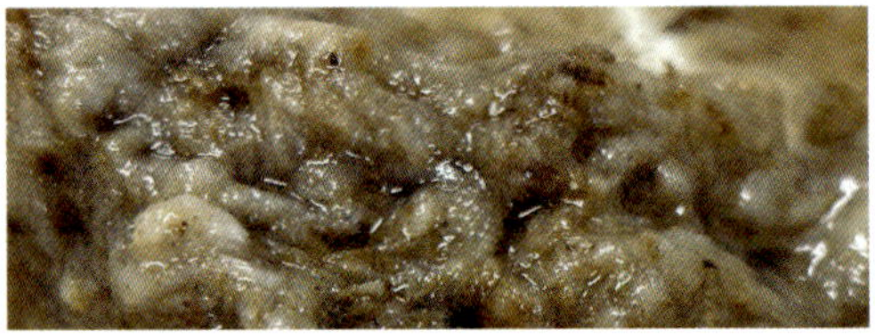

Photo credits: Miles Monson, Kui Hu, Adam Heathcote, and Mark Edlund from Science Museum of Minnesota

Q: DO ROCKS CATCH COLDS?

A: It's true rock snot looks like the stuff that comes out of a runny nose. But no, rocks do not catch colds. Rock snot doesn't actually feel slippery to your paw or hand—it feels like a wet wool sweater!

Q: WHAT DOES ROCK SNOT LOOK LIKE?

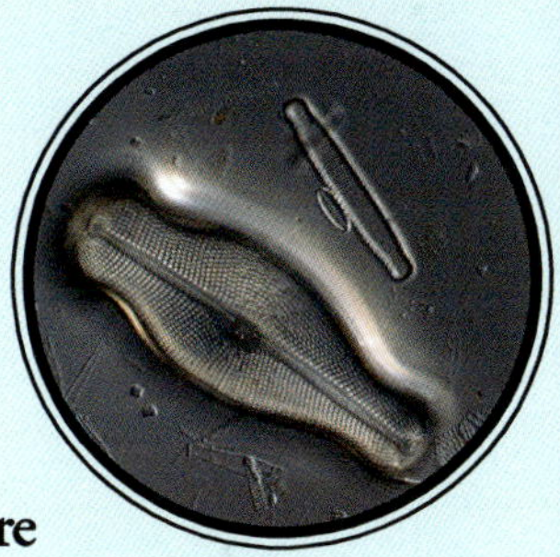

A: Rock snot's tiny cells are invisible unless you have a fancy microscope like scientists do to magnify the rock snot cell 1,000 times its natural size.

But rock snot cells grow long golden brown stalks that are easy to see without a microscope. That's what you see covering rocks—stalks of long gloppy, snotty, yucky mess!

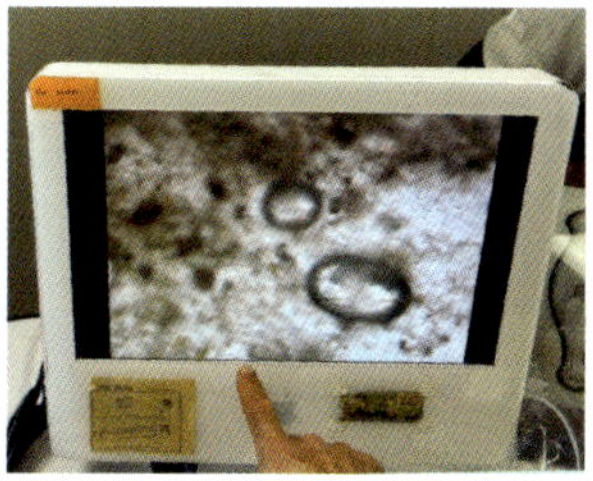

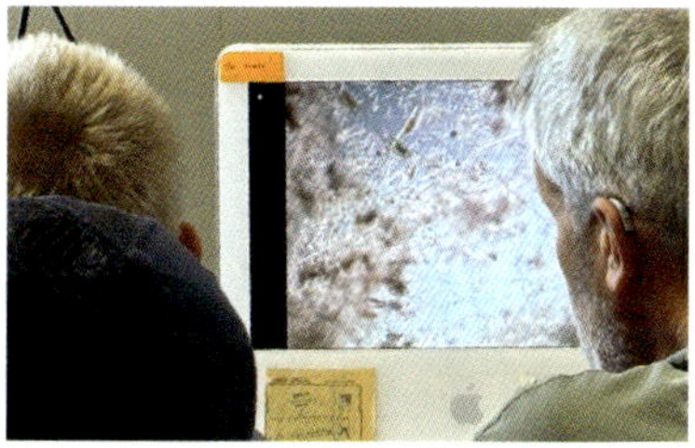

Art by Kip Hathaway; Photo credits: Science Museum of Minnesota, Anne Brataas

Q: WHERE DID ROCK SNOT COME FROM?

A: No one is sure. It is found in cold water lakes around the world. Scientists think rock snot is a natural part of those lakes. In the past, it has been in such low levels that it doesn't degrade water or disrupt fish or insect habitat.

The first written report about the discovery of rock snot in the world was by a Danish scientist more than 100 years ago. He found rock snot in a Greenland lake. In Minnesota, Lake Superior rock snot was first reported in 1961. But it didn't seem to be causing problems. Now rock snot seems to be moving into waters around Lake Superior. Each year, teams of scientists test approximately 25 North Shore streams to monitor the spread of rock snot.

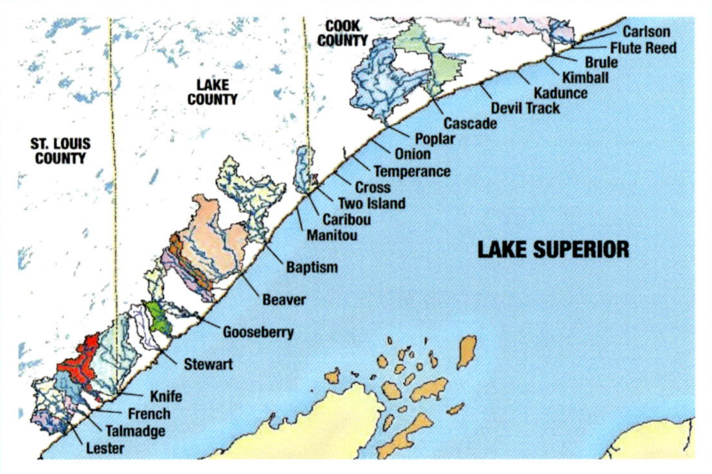

Map of North Shore streams tested for rock snot since 2018

Map credit: Minnesota Department of Natural Resources

Q: IS ROCK SNOT SPREADING IN MINNESOTA'S NORTH SHORE WATERS?

A: Yes. Something is changing in Minnesota waters to promote the growth of rock snot. In 2018, scientists from the Science Museum of Minnesota's St. Croix Watershed Research Station teamed up with other scientists from the Minnesota Department of Natural Resources and University of Minnesota-Duluth to sample North Shore streams and lakes for rock snot. They first found rock snot in the Poplar River by Lutsen, Minnesota!

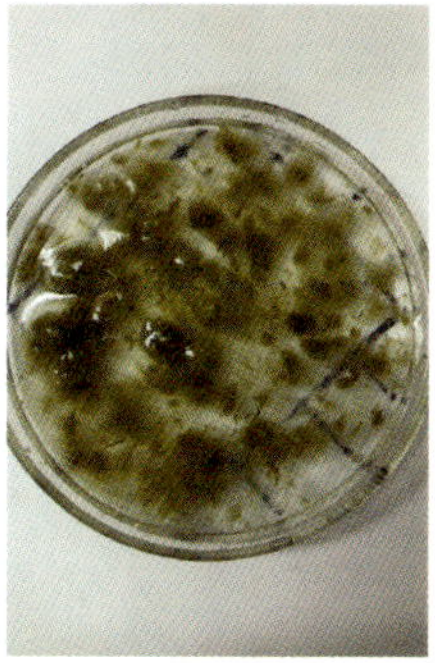

Top: Underwater photos of rock snot covering rocks on a stream bottom

Middle: Scientists collecting samples of fish, insects, and water on the North Shore to test for rock snot

Bottom: Researchers analyzing water samples at the St. Croix Watershed Research Station laboratory

Photo credits: Minnesota Department of Natural Resources and Science Museum of Minnesota

Scientists are investigating this question:

"What is changing to allow rock snot to thrive and spread?"

Their **HYPOTHESES** involve effects of climate change and increased human use of the North Shore, prompting questions such as:

- Is the chemistry of the water different?
- Is the water temperature warmer?
- Are mutations in the genetics of rock snot helping it adapt to North Shore waters?
- Are fish diets changing?
- How are insects, bacteria and other tiny plants, animals and microscopic organisms affected by rock snot?

Scientists testing Lake Superior and North Shore streams

Photo credits: David Burge and Mari Leland from the Science Museum of Minnesota

Q: ARE SCIENTISTS WORKING WITH YOUTH IN MINNESOTA TO HELP INCREASE PUBLIC AWARENESS ABOUT ROCK SNOT?

A: Yes! Scientists are working with Story Scouts in Grand Marais, Minnesota, along the North Shore to teach kids about the science of rock snot, and how to take water samples and identify rock snot under the microscope. Kids used their scientific understanding to write and illustrate this book and tell people how to protect water and all that lives in it.

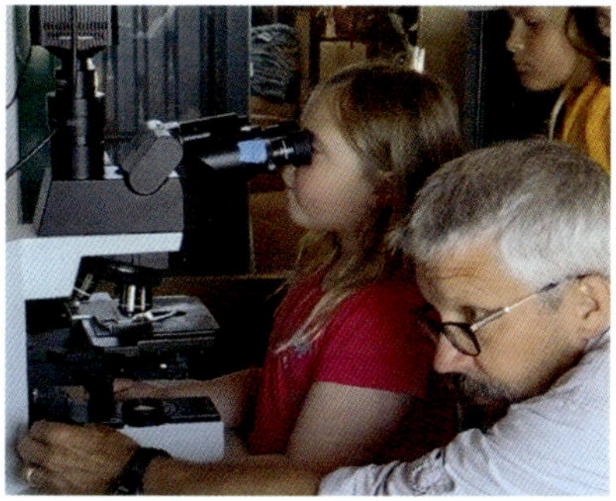

Photo credits: Anne Brataas

Q: HOW DO SCIENTISTS AND YOUTH WORK TOGETHER TO TEACH PEOPLE ABOUT ROCK SNOT?

A: They collaborate and color! In 2022, a Minnesota Department of Natural Resources (DNR) scientist made coloring sheets to help people visualize the threat rock snot poses to water life. Then, in 2024, Story Scouts made simpler versions for young kids. Download and color them yourselves to see how cold and clean water that is FREE of rock snot looks compared to how messy and dirty water FULL of rock snot looks.

Coloring sheet PDFs are free to download at www.rocksnot.org.

WATER FREE OF ROCK SNOT

Art by Maite, Brisa, Sam, and Iona

WATER FULL OF ROCK SNOT

1. Has lots of stalks. Looks messy.
2. Disrupts healthy water habitat.
3. Smothers insect larvae that fish eat to survive.
4. Lack of a good insect food source may cause fish to migrate to better feeding grounds, or cause them to grow more slowly, and be fewer in number.

Art by Freja, Aren, Selah, and Iona;
Coloring sheet credit: based on a Minnesota Department of Natural Resources illustration by Danielle Kuball ©2022

Q: DOES ROCK SNOT HAVE ANY NATURAL PREDATORS TO CONTROL IT?

A: No!!! And that's the problem. We wish there were predators! Right now, the only way to stop the spread of rock snot is for people to change their habits when they move between bodies of water.

People need to clean their boots, boats, and gear when they leave one body of water and before entering the next.

While waiting for scientists to discover ways to control rock snot, Story Scouts have invented possible predators called ROCK SNOTTERS.

ROCK SNOTTERS

Art by Ethan G., George, and Sig

Q: HOW DO WE GET RID OF ROCK SNOT AND PROTECT FRESH WATER?

A: By caring and being careful not to "track" rock snot bits stuck on boots, boats, or gear such as fishing poles, lures, paddles, and oars when we move between lakes and streams.

If we don't wash our fur or gear off after we've been in a lake, river, or stream, we could track rock snot into the next lake or river we enter. And then it could make rock snot! Eeeeeeewwwwwwww!!!

Art by Mikey

After we watched the Q&A video, we thanked our new fish friends and swam to shore with the tablet. Some of us are retriever puppies, after all.

Dr. Definitely was there—she was so busy talking to the scientists who lost their tablet that she seemed to forget she told us not to go swimming.

We apologized to her for not following directions. Dr. Definitely forgave us because we were really sorry. You could tell by our big sad puppy eyes.

And the scientists were happy because we got their tablet back to them!

Art by Grace (rainbow cave), Rilynn (fish and puppies saying farewell), and Brynn (sad pup)

The scientists gave us a bath in a plastic tub they carried in their truck just to be extra careful we wouldn't spread rock snot. Using hot soapy water, they rinsed our fur, bandanas, collars, backpacks, and hoodies. Then they poured the tub water into a jug and took it home to water their gardens.

We were clean and shiny and happy as fish—so happy we ran in circles again!

Dog art by Jocelyn, Runa, Rosalynn, Sabrina, Tib, Freja, Nolan, and Leah

PART 4:

OUR IMAGINARY SOLUTION: Rock Snotters

Luckily, scientists are working on discovering and developing better ways to stop rock snot from invading our waters and spreading.

But face it: this is a job for puppies! To be precise, the puppies of Sunny Shores Puppy Preschool!

We invented a creature we think could be a super-duper predator of rock snot. We call it a Rock Snotter. Rock Snotters drink rock snot, eat Rock Snotty Joes at school lunch, and make super chewy candy out of rock snot: Sour Snotties.

You can see them in our picture gallery, which we call **Revenge of the Rock Snotters Hall of Awesomeness**.

Take a look! And if it gives you ideas of how to control rock snot, please tell us about it, draw a picture, and send it to us!

REVENGE OF THE ROCK SNOTTERS HALL OF AWESOMENESS

Art by (clockwise from top left) Ethan A., Ethan G., Anson, and Leah

REVENGE OF THE ROCK SNOTTERS HALL OF AWESOMENESS

Art by Mavis

Dear Readers and Wonderers:

Please add your own pictures to our Hall of Awesomeness! If you have ideas about how to control rock snot, please draw and write about them.

Ask a grown-up to take a photo of your creation and email it to us: rocksnot@minnchildpress.org. We will proudly display it on our website gallery at www.minnchildpress/rocksnot.org.

This is one way we can help scientists use their imaginations—like puppies and children naturally do!—to get new ideas on how to stop the spread of rock snot.

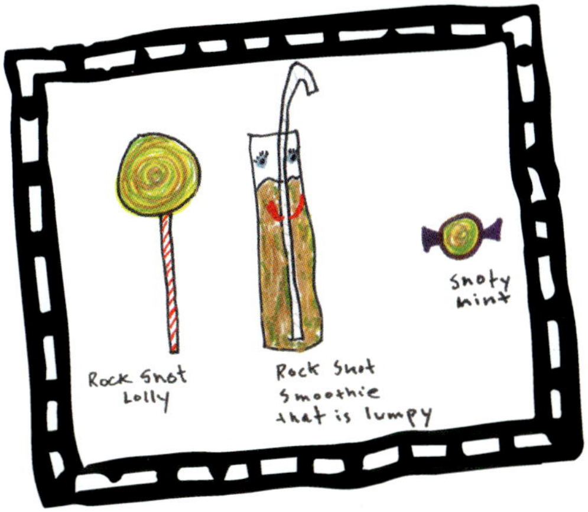

Draw your art here

Art by (top to bottom) Emily and Ella P.

MORE AWESOME STRATEGIES

Here are a few ideas we think scientists should consider to control rock snot.

1. DISGUISE

At "Teacher Dress-up Day" Dr. Definitely shocked students with this outfit so much that they ran out of school. Could we shock rock snot out of the water with a disguise? Giving it the evil eye?

2. MERMAID MONITORS

An alarm system—how about mermaids!—could alert scientists when rock snot starts growing so scientists can remove it before it takes over.

Art (top) by Elisa (evil eyes) and Adalyn (fancy dog); (bottom mermaids L-R) by Quinn, Selah, Brisa, and Maite

3. MEDIEVAL WARFARE

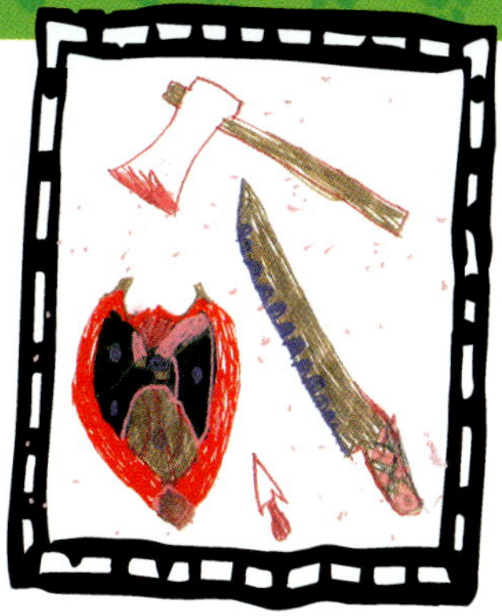

Maybe there's a machine to shield fish and larvae from rock snot, or to attack it with axes, blades, and blades that turn into combat fish.

5. TICKLING

It's worth trying tickling rock snot into submission. A feather would be a handy tool for that.

6. GOLDEN UNICORN

Their shiny magic might protect fish and insect larvae from rock snot.

7. PUPPIES

Everything's better with puppies, including puppies with rainbow wings!

Art (clockwise from top middle) by Elliott, Dax, Freja, Sabrina, Freddie, and Felix

For more ideas on how to control rock snot, invent your own Hackers & Hoodies to get inspired!

"Circle on" art by Ari, Brisa, Brynn, Freja, Grace, Jocelyn, Leah, Mavis, Rilynn, Rosalynn, Runa, Sabrina, Selah, Tib, and Kip Hathaway

Invite us to join you! That way we can run and swim circles together! Circle on!

THE END

BONUS!

SCIENCE WORDS AND CONCEPTS

ALGAE are simple, ancient life forms found naturally in lakes, rivers, and streams that looks like a plant and are often called "sea weed." There are thousands of type of algae, and most are important sources of food for aquatic organisms. Some have poisons that can make dogs or people sick if they drink water polluted by it.

AQUATIC is a descriptive word for watery things. An "aquatic environment" could mean a lake, pond, stream, and an "aquatic organism" is a water life form.

BIOLOGY is the science of living things.

BODY OF WATER is a general term that refers to any natural surface water contained by a landscape feature such as a basin (lakes, ponds) or bed (rivers, streams).

DIATOM is a form of algae distinctive for containing a glass-like cell wall made of silica, a hard mineral found in quartz and dissolved in our waters that can help protect diatoms from predators.

DIDYMOSPHENIA GEMINATA (didymo, or rock snot for short) is the formal scientific name given to rock snot more than 100 years ago in the Latin language. Part of its name (geminata) more likely refers to the twin cells at the end of a goopy, slimy branch: when magnified, a single rock snot cell resembles a double-headed bowling pin—it has twin knobby ends.

ECOLOGY is the science that studies and describes relationships between living things. Rock snot threatens to disrupt stream ecology by forming thick mats on stream beds, which can displace or kill aquatic organisms such as insect larvae that live there and that fish need to eat.

ECOSYSTEM refers to the interactive biological networks of a place that sustain life—or threaten it when conditions change.

FANCY MICROSCOPE is a key piece of lab technology needed to magnify microscopic rock snot cells so scientists can analyze them, identify them, and see if rock snot species are changing one stream to the next, and from year to year. This will help scientists understand rock snot and design systems for eliminating or controlling it to protect North Shore streams. The Science Museum of Minnesota researchers use a compound light microscope that magnifies the specimen 1000 times, illuminated with DIC oil immersion and PlanApo lens with a numerical aperture of 1.4. In other words, it has awesome optics!

HABITAT refers to the natural environment a living organism occupies.

HACKERS & HOODIES is not really a science term, unless you think of hackers as using computer science to disrupt intended functions of computers. Mainly it's part of a store name Story Scouts made up!

HYPOTHESES are scientific explanations based on evidence and observations that are tested to see if they are valid. As more evidence is discovered, hypotheses can change and be tested again.

INSECT LARVAE are the young forms of animals such as insects and butterflies that look very different from their adult forms. As larvae grow, they change by metamorphosis into winged adults. Many aquatic insects have a worm-like larval stage in water. Adults emerge into the air and land where they can fly.

PREDATOR is a role in animal food-chain relationships that hunts, kills, and eats other organisms ("prey") to obtain nutrition and survive.

ROCK SNOT is the nickname for the invasive diatom that overgrows northern streambeds to disrupt healthy habitat, potentially threatening fish life and spoiling recreational experiences such as fishing, boating, and swimming.

ROCK SNOTTER is an imaginary predator that Story Scouts made up to eat rock snot and clean up water polluted with it.

STREAM is used as a general term for natural waters, whether they are lakes, ponds, streams, or rivers, as in "North Shore streams." It is also a specific term for a channel of running water contained by a geologic landform known as a streambed. A stream differs from a river by its size; streams are smaller and shorter and drain a smaller watershed than rivers.

THE BIRTH OF SPARKY

BEFORE

AFTER

Art by Ari; Photo credit: Anne Brataas

Hi
Hi
Hi
Hi